D0742003

Dudleya Plant, Upper Baja

The Glory of Nature's Form

The Glory of Nature's Form

photography and text
by
Willis Peterson

Dedicated to all my students—past, present and future.

Mountain Lion, California

Publisher: Robert D. Shangle
Beautiful America Publishing Company
Beaverton, Oregon 97005

Design: Cynthia Peterson

First Printing July, 1979

Library of Congress Cataloging in Publication Data

Peterson, Willis.
The Glory of Nature's Form.

1. Natural history—Pictorial works. I. Title.
QH46.P47 500.9'02'22 79-12418
ISBN 0-89802-001-8

Sunrise over Monument Valley, Arizona

Contents

Seeing the Essence

All lovely in their own way, plants and wild creatures create a living world of infinite color, mood and design. The shyness of a fawn, the indescribable hue of a flower, the points of light refracting from droplets of dew are but ramifications of a far greater experience—the seeing of nature through a camera. To register on film such subjective nuances is an exhilarating, spiritual experience for me.

There is more involved than just the mechanical recording of an image through a lens. It becomes a three-way alchemy of a sophisticated technical skill, an innate desire to make photographs under any conditions, and a sense of knowing when to portray the mood and essence behind the physical likeness in the image.

Perhaps this last point is the most important.

The image must have a vibrant composition so it will live. It has to inspire a thought. It has to communicate a statement, that this is spring, or this is fall, or this is love, or this is youth, etc. The nature picture must show this mood with the subject as its star performer.

Along with portraying the mood, there is a second and equally important consideration. It is the sense of discovering the spirit or essence associated with the subject's physical being. Hard to define? Maybe, but it will come with photographic experience and it will come with seeing and feeling the beauty of life's images.

The unique interlude I had while photographing the Alaskan Dall Sheep reveals my sentiment more clearly. Certainly, their horns sweep into a full curl, their heads are magnificent to behold, their stance upon the crags defies description.

My goal as a nature photographer is to portray these physical traits as a living spirit. Then, does the subject assume that magical, spiritual essence of life. Combined with the mood of the photograph we then see past the likeness into something more significant.

I release the shutter only when I feel this essence, this truth to be at its highest degree. My camera, my lens, my desire, and my judgement of when to shoot, interact to portray and describe the subject in image form as an eternal truth.

Only then does the image portray "I am the majestic Dall, prince of the crags."

This is my credo of nature photography—defining in an image the "essence" of life.

Dall Sheep, Alaska

The Mountain Vistas

A wild and vibrant tableau of mountain peaks floods our senses with stimulating and inspiring emotion.

Mt. Rainier, Washington

The mountains have always been very special to me. There has scarcely been a day since I was small that I haven't gazed some time during the day upon those ponderous ramparts of nature's extravagance.

I can still remember with utmost clarity my first glimpse of the Rockies. I was six then and filled with awe, as we entered Eastern Colorado on the Rock Island Line. When the conductor entered our swaying coach and imperiously announced that the Rockies were visible, I crowded to the window of the Pullman to see this wondrous apparition. A thin ribbon of dark blue had appeared across the far western horizon.

While the train high-balled over the Plains, I stared with impatience at that growing and foreboding chain of mountains. Little by little distant shoulders of blue became more distinct. I was thrilled beyond words when Pike's Peak, with its upper half mantled in brilliant white, emerged from a shroud of clouds.

At the railroad station in Colorado Springs I scrambled from the car's vestibule while grimly hanging on to my aunt's skirt. In front of me a few miles away the whole Rampart Range towered to heavenly proportions. Fragrance of spruce and fir flooded the air. I drank in the aromatic air and stared at the mountains. To top it all off, Ute Indians, wearing all their finery, brushed by us. This was to be my country.

Even now the mountain vistas offer an appealing charisma to me.

Mountains provide a stark reality. They provide an identity for people. It is no wonder that mountaineers have always been regarded as such rugged individualists. This independent quality seems to be apparent even in the animals and plants which live upon the mountains.

I recall, one time, as my son and I were hiking at 11,000 feet in the White Mountains of California when this trait became so evident. We had slipped our pack boards momentarily and were looking down into the great Owens Valley. Tiny farms and cities so far below appeared as though they were plastic models. Across the Valley another wave of mountains reached skyward, the mighty Sierras.

To get a better view we continued to climb, picking our way carefully across the jumbled scree, when suddenly we came upon a grove of those oldest of life forms, the bristlecone pines. They existed upon the very steepest, stoniest of peaks and seemed to be in a continual battle with the elements. Their grizzled personalities had endured eons of time. Thousands of years had elapsed since those gnarled forms had been seedlings.

One of the pines seemed particularly grotesque, yet beautiful. Reaching, supplicating, seeming to cry in agony, the wind-burned limbs of the ancient tree struggled skyward in a wierd embrace.

It projected such an intimate feeling of preserverance that at first we could only stare. Then, concentrating on the angle of the tree's trunk, I squirmed into a low position to catch its agonized existence in my camera.

I rearranged the composition to include a tiny bit more of the striated colors. Browns and yellows, glowing in woody warmth, met my eye. Textures stretched tenuously along the scrolls of weathered grain. Showing delicately, they exuded the tenacity that still imbued this living thing. The rest of the twisted trunk showed in detail. It's 4,000 year old excruciating life still defied the elements. Only the mountains could have spawned such an individual.

I think everybody has their favorite mountain range. Of course, the Rockies created their own fascination with me those many years ago. In particular, Pike's Peak, the age-old patriarch surrounded by the Rampart Range was my first mountain love. It was so gigantic and I was so small that I felt engulfed by its magnitude. I was drawn to its snowy facade. I was sure it had magical power.

When I saw Mt. Rainier for the first time I thought that surely here was the loveliest of all American peaks. I still do. It is brilliantly mantled in ermine even in summer. Lower slopes are carpeted with fields of wildflowers, giving it a dream-like appearance that can never be forgotten.

For sheer mountain majesty Mt. McKinley is unsurpassed. The Indians, who were so much attuned to their environment, called the Mountain, Denali, "The High One." How simply stated. How dramatically apropos. They revered it as a manifestation of their Great Spirit. In the predawn light, Denali's towering, icy pillars and massive, snowy summits have an unearthly look of moonglow. As the sun nears the horizon, the colossus takes on a faint pink cast which slowly envelops the Mountain's flanks. Finally, overwhelming one's senses with awe, a salmon-pink citadel soaring 20,320 feet in the sky reveals itself.

The austere but magnificent desert mountains offer similar reflective, dramatic moods. During late afternoons when lengthening shadows silhouette their jagged heights, the feel of solitude is often enhanced. This is the time when I like to climb to a crest. I can see for miles across the low valleys and, in fact, on a clear evening I can pick out a flicker of a candle a mile away.

The most spectacular time of day occurs at sundown. There always seems to be a magical aura about sundown as it transcends over the desert mountains. It is the portion of day when light and design of nature become as one. The result is a kaleidoscope of shapes and forms, all in stark silhouette, yet melting into one another, receding before one's eyes into a sense of timelessness.

Pikes Peak, Colorado

Mountain Stream, California

Mt. Rainier Reflection, Washington
Black Bear, Wyoming

Pike's Peak, Colorado

Rocky Mountains, Colorado

Elk, Wyoming

Grand Tetons, Jackson Lake, Wyoming

Bristlecone Pine, California

Yellow Columbine, Colorado

High Sierra Mountains, California

Hanging Lake, Colorado

Moun ain Backdrop, Washington

The Northland Reaches

A land of ice and snow punctuated yearly by 20 hour days and vivid greens make up a wondrous Arctic summer.

Moose in Wonder Lake, Alaska

I scanned the horizon. Mountains slanted skyward, stark and somber, while jutting from the sea, forested islands followed along the coast. Northward, beyond the ocean eddies and mists that swirled about the promentories, lay the historic gold fields that had lured adventurers and the '98'ers and opened up a land so majestic that it staggered the imagination.

Those ermine clad peaks, those glistening sheets of ice creeping down from a horizon of perpetual snow, those summers where the land never sleeps, those reaches of tundra teeming with life, those creatures, the bear and the wolf, the caribou, the moose, those birds, those flowers, those stupendous panoramas leading on and on—fused into a single nostalgic wonderment. Alaska. Surely, such an image was gold in itself.

While countless gold seekers had come and gone it was the Robert Services and the Jack Londons who are the remembered ones in that hectic era of the "Rush." The poets saw the romance and beauty of the country.

It was with a similar fascination of the Northlands that I vowed to experience. I wanted to see if through the camera I could also portray the Northland as poetic imagery.

The landscape reflected and dazzled me with brilliant light. The Arctic summer sun, there is nothing quite like it—so prismatic—it bore deep into the viable green of the tundra, backlighting a million living sprigs into olive-green prisms.

Shining ground cherry bushes cradled to their bosom clusters of red berries. Low bush cranberries were prostrate, too, in the same pregnant fulfillment of summer's gift. And the blueberry bushes, loaded with dusky fruit, swept across the verdant breast of the tundra. With camera slung over my shoulder, I plunged ahead through the Northland tangle of green.

Ahead of me each bit of green blended with more greens. Solid folds of verdure parted only where ravines and canyons led down to naked, glacial outwashes and turbulent, glacial streams. The whole promise of the land resolved into a single purpose—to recreate itself.

I followed a gurling freshet through waves of tundra. On either bank, pillows of chartreuse mosses guided and held captive the glistening, trickling water until it tumbled into a shinging pond. Shown in crystaled sheen, the lake was framed by the fireweed, that prolific, magenta flowered wanderer of the Northland.

In the fall all this would suddenly change. As if touched by a magician's wand, the tundra would turn into beautiful, intricate colored patterns.

From the Kenai I photographed Portage Glacier. Spilling down from the Chugach Mountains, Portage is classic in confirmation, beautiful in coloration. Its pale blue ice wall calved into beautifully shaped, sculptured bergs. Often driven across the lake by fierce winds, the bergs would ground ashore and remain there like huge chessmen. As each crystallined edge caught the light, it revealed in my viewfinder an infinite number of tiny stars.

I camped upon the terminal moraine of Mendenhall Glacier. From the crest of the gravel deposits, I gasped in amazement at that formidable barrier of blue that crept so slowly, so steadily, so relentlessly down from the icefields.

"How old are you?" I mused. ' What was America when you were born? When did you fall from the sky in crystals so lacy cold? How long have you traveled from your endless vaults of snow?"

Packed and highly compressed, the wall was more than one hundred feet high and carved into a fantastic variety of fissured scallops. A particularly blue gouge revealed where a recent calve had crumbled into the lake. One fell during the night, and though a good mile away, the cracking, rumble and final report startled me out of my sleep. The next day I saw where it had fallen. I judged one thousand tons of ice had been hurled into the lake.

Below me, the bold face of the glacier lay mirrored in water so still that there appeared to be two icy barriers. Its inverted symmetry was distracting, its resplendent perfection disturbing, for it seemed that anything so exquisitely engraved upon the land could scarcely emanate from forces so monstrous. Caught in the onslaught of that crushing ice, a house-size boulder had no more tranquility of residence than an uprooted garden bulb.

My most memorable encounter with wildlife was photographing the Alaskan dall sheep.

Those intrepid mountaineers—they personified the spirit of freedom. They and their environment were exponents of this precious commodity. Their life on the pinnacles and peaks offered no other choice. They needed the vastness of plummeting space and sheer mountain backbones to make their life complete.

Now, after a backbreaking hike I was within 200 feet of fifty-seven rams, huge rams, rams in immaculate white, rams with flaring golden horns reflecting the sunlight. While a few sentinels watched the valley most lounged on the carpet of tundra. Their white forms drawn against velvet green, with stands of wildflowers adding vivid spots of color, created an astounding landscape.

I edged closer and closer, and still the rams did not regard me with fear. Some lay in shallow beds they had kicked out from loose shale. Some were surrounded by flowers. Others rested massive heads upon the tundra. A few stared off into space, scrutinizing some faraway form of life only they could see.

It seemed unbelievable until I squeezed off the first picture frame. It was not a dream after all.

Iceberg, Portage Glacier, Alaska

Trumpeter Swan, Kenai Peninsula

Tundra Ponds, Alaska

Mendenhall Glacier, Alaska

Caribou, Alaska

Grizzly Bear, Yukon

Hanging Glacier, Columbia Ice Fields, Canada

Fall on the Tundra, Alaska

Mt. McKinley, Alaska

Marmot, Alaska

Lake in Canadian Rockies, Canada

The Coastal Strands

Where land and sea meet,
a fascinating world of tides, dunes and
shores create a picturesque landscape.

Coastal Dunes, Oregon

The meeting of the land and the sea, where the tryst of wave and strand create an infinite exchange of moods, does strange things to my consciousness. The coolness, the spray, the beach, the fog brings on a strange euphoria.

The unhesitating sea breaking over the rocks and then the water sluicing away again and again provide as many different compositions as there are waves. I expect this is the challenging expectancy of making pictures of the sea—to portray this continual changing mood.

From Baja to the Aleutian Chain, this pulse, this ceaseless rhythm of the globe itself continues relentlessly. The cresting wave, the gentle swell, the sheen of light on water in the evening are all ramifications of this eternal pulse. It is this tempo that the photographer must interpret.

Some of the most spectacular sunsets I've seen have been when this pulse is at its most quiescent mood. There are times when the whole sky over the Sea of Cortez literally turns to crimson. The warmth of the tropical sky is complimented and radiated across the shimmering sea.

Evenings over the North Pacific are different. Many times only a thin band along the horizon changes into lavendar and then lingers coldly against the sea. Shadows between the shore dunes pick up the color and transmit to the viewer a cold, magenta-tinted blue.

In the foreground the sea grasses bow with the chilling wind. Tussock after tussock recedes before the camera, all bending and facing the same way. Holding tenaciously onto a firmament, they fight for every inch of sand that the ocean will allow. In the background, sand, sky and sea come together in a cold crucible of night.

Even so, life prevails. The sea otter, whimsical, playful, once so plentiful along our Northwestern coast lives in the surf and tidal pools. They feed upon abalone or small squid. When they are finished with diving they roll over and place their newly gotten dinner upon their chest. While floating in this blissful manner, they leisurely break the shell open with a rock and pick out the meat.

One of the strangest creatures of the sea found along the Pacific Coast is the sea elephant. The bulls haul out upon the beaches in early spring to gather harems. Their huge proboscis and monstrous bodies are often covered with scars from enumerable battles garnered during this strenuous season. Flippers are used for locomotion. Their "hauling out" is seemingly an agonizing effort when you consider a bull may weigh a ton.

Point Lobos is still known for its sea lions. Their continuous baying reminded the early Spanish explorers of howling wolves. Taking advantage of the shelters along the coast, the sea lions clamber upon the rocks to bask in the sun. Their young are born on land, and strangely, must be taught to swim. The tidal pools become crowded nurseries during this busy time.

One of my greatest photographic moments occurred when I spotted a pilot whale. I followed it with my longest telephoto lens. When it finally appeared in the viewfinder in a full broadside position I started shooting. Then wonder of wonders, the whale blew. Steam, spray and mist shot from the blow hole. The vapor spouted upward and then fell in a filamentous curtain of spray. At that precise moment a rainbow appeared in the enveloping mist.

I've often thought about that shot of shots. It was a happy accident, yet I was prepared. The right light angle, the precise aperture setting, the correct slant of the sun, the right shutter speed all combined to show that essence of spirit.

The Oregon Coast provides some of the most picturesque scenes on the Pacific. But to comprehend the majesty of the sea meeting the land one must climb to a commanding height. Then the curvature of the waves thrusting against the lonely strands combine to make an image of stupendous proportions. The elevation creates a sense of slow motion as the ocean's pulse continues far below. The mysterious feel of the sea is intensified as the fog and mist envelope the distant headlands.

Along the Alaskan coast, on the days that the elements would slacken their pace, I would often walk out upon the sea cliffs and perch upon the gnarled and twisted tree roots that had long since lost their life to the devouring sea.

Before me the fog, driven by myriad vexing air currents, swirled across the Cook Inlet. The capricious mist was in sharp contrast with the change in the tide that rose irrestibly, relentlessly, forcing wave after wave upon the beach in ever increasing volumes until it inundated the barren flats.

When it turned, it ran out to sea again just as violently. And nothing could stop it, as if having a tryst with the land, the tide had to replenish its strength with an unfathomable sea far away.

Waters receded rapidly and where just a few minutes ago they were lapping along the shore, they were now only found in the main channel. Gulls wheeled overhead to search for offerings, which the tide in its benevolence had left behind. They alighted on the newly glistening spits of suddenly denuded, muddy flats and fought among themselves over an old salmon that had made its last journey, or some other dead citizen of the deep. Mists, rain and cloud suddenly closed in, clothing the ribald scene, and hid the gristly feed.

On the beach, nothing can be as fascinating for the photographer as the little bits of flotsam found along the wave line. Chaotic heaps of seaweeds, piles of bleaching driftwood cast up from the hurrying tide offer countless images for the photographer to explore. Amidst this somber perspective of nature, crabs scuttling across the strand can reward a photographer with humorous sidelights.

Waves, Pacific Coast

Driftwood in Ice Plants, Point Lobos, California

Forested Headlands, Oregon

Bahia Concepcion, Lower Baja

Fiddler Crab, Lower Baja

Oregon Coast

Sunset, Sea of Cortez

Pilot Whale, Upper Baja

Sea Elephant, California

Sea Otter, California

Palm Trees, San Blas, Mexico

The Desertland Lure

Sometimes conservative, sometimes flamboyant, the austere desert demands that its citizens adapt to many whims.

Evening Primrose, Coral Dunes, Utah

After living in the shadow of the Rockies for so many years, the move I made to the deserts of Arizona seemed foreign, indeed. As a matter of fact, I am still adapting to the desert.

Instead of spruce and fir, saguaros and cholla stood, rather than air perfumed with conifers there was the pungent smell of creosote bushes. Gray, stickery shrubs substituted for the luxuriant green plants of the forest. Nothing could have been so dramatically different.

Even at that, the connotation of the word ''desert'' is in a way misleading, for it implies a barren, uninhabitable land.

On the contrary, I found the desert to be a living thing. Its kaleidoscope of life forms from tiny ferns to cactuses including a prolific number of birds and animals indicate that it may not be such a wasteland at that. The desert is always renewed quickly with a little water. Even man, with his ingenuity and manipulation, has made gardens out of deserts past, since the time of Babylon and before.

Granted, it must lie dormant at times. Those are the times when its citizens must move about sparingly, and when its plants must remain conservative. But, ''desert'', has such an air of finality. It is an interpretation, which is too definitive for my land of sun. The desert may restrict life according to its whims, but on the other hand, it is quite flamboyant under the right conditions.

Little by little I learned these desert ways. I frequented the washes and discovered that the tracks in the sandy bottoms were proof that a veritable multitude of animals and birds did live and prosper in this seemingly land of thorny hostility.

Prints made by these creatures revealed a great deal about desert life. A flurry of scratches here and there meant that rodents and birds, too, used the sand as a gritty, cleansing tub. Fine, stitch-like tracks told of insects marching to better food plants. Heavy gouges indicated where predator and victim had had a fatal confrontation. Large, round indentations revealed where a pair of bobcats had strolled. Sharp, deep prints meant that javalina and deer had passed by.

The desert is host to many picturesque creatures. The Gambel's quail with their cheerful calling are regular Beau Brummells of the avain population. The road runner, too, with his strange and erratic antics provides the desert with a comic relief. The ringtail cat, coati mundi and kangaroo rat are fascinating to observe. The latter has adapted to the desert so well that it can manufacture water from his own metabolism. It need not drink at all!

Aside from discovering such abundant animal life, I think the blooming of the annuals is one of the most wondrous occurences I've ever had the pleasure to witness. Seemingly, springing from nowhere, tiny seeds germinate, grow and cover the desert with buds. Suddenly as though a magician should pass a wand, plants of all descriptions are covered with flowers. Such an array of chromatic enchantment follows that it bewilders the senses. Where only months before nothing except desicated shrubs, sand and rocks appeared, the whole landscape has now been transformed into a wonderland.

Shimmering fields of gold poppies give the impression of flowing, molten metal, while mixed in between, stands of lupine form an exciting sky-blue contrast. Magenta hued splotches of owl clover creep under green speared ocotillos. Tiny goldfields bloom in solid mats beneath stands of saguaros.

The bulbs and perennial herbs produce just as colorful array. The Ajo lily, stately white, exquisitely proportioned, stands demurely with bowed crown. If the preceeding springtimes have been dry, five or six years may have elapsed since this plant last ventured toward the sun.

Later in the season, mariposa lilies unfold their orange-red petals. Aptly named by the early Spanish, ''mariposa'' means butterfly. Honeysuckle-like flower tubes of the hummingbird bush create billowing clouds of red along the washes, while everywhere, wands of apricot-colored mallow wave to and fro in the desert breeze.

Its luring magic is overwhelming. In the infinite wisdom of nature, these seeds have lain dormant for months, some, for years. But, as winter rains fall, chemical inhibitors have been soaked away from the seed coatings. In turn, the tiny embryo of life has been unlocked. Strangely enough, only the right combination of rainfall and temperature will affect it. If not, the seed doesn't respond and remains at rest.

Variance of temperature and moisture, linked with physical characteristics of the terrain are chief factors which isolate group species from each other. Vegetative cover, soil minerals, drainage and combination of all these forces produce the multi-colored floral displays. Once witnessed, they surely aren't forgotten. In fact, I can recall all the prolific blooming years since I came to live upon the desert some thirty-five years ago.

As to time of day, I've always thought that the most stirring time to view the desert is the late afternoon. This is the hour when the light striking the needles of the cactus plants, especially the chollas, creates such striking halos, accentuating and boldly outlining their trunks and branches. As shadows lengthen fascinating perspectives form. An enchanted dimensional world of stark shapes meets the eye. The rest of the backlighted landscape is sharply etched in highlight and shadow.

As the sun sinks lower the sky turns to red-orange color. The view seems to lure the visitor irresistibly before the red wash fades away.

Owl Clover and California Poppies, Arizona

Organ Pipe National Monument, Arizona

Calfcreek Falls, Utah

Coral Dunes, Utah

Baja Cactus, Lower Baja

Gold Poppies, Arizona

Pichacho Peak and Poppies, Arizona

Kangaroo Rat, Arizona

Desert Bighorn Sheep, Nevada

Sunset, Utah

Delicate Arch, Utah

Landscape Arch, Utah

Canyon de Chelly, Arizona

Chisos Mts., Texas

Cambel Quail, Arizona

The Plains Infinity

The open prairie and grasslands leading on and on have always attracted the hardy and fleet of foot.

Badlands, South Dakota

While the mountains repose in grand eloquence and form a finite landscape to the eye, it is the Plains that bring to one a feeling of infinity.

The undulating landscape, slightly rising now, the dipping down into swales, forces the wanderer to go ever onward, to seek and to search and to ultimately discover the meaning to a trackless expanse.

When I was a boy I would often hike into the vast stretches of grassland that bordered against the Rockies. I would walk for miles over the terrain which would seem to be endless and the same, yet there would be dramatic changes.

In the spring where water had collected into shallow ponds, the new grasses would be showing under the cover of the old. The cowboys with their wind-burned faces and crinkled eyes would tell me that these depressions were once buffalo wallows where, years ago, these immense creatures rolled and tore at the sod with their horns. Now the wallows were silent, only shallows where occasional violets bloomed demurely along the edges.

Near the ponds killdeer would skitter ahead of me, calling, always calling in their plaintive manner. Horned larks would suddenly run out from clumps of grass and prostrate bushes.

In late summer the colors were mostly amber as though the Plains were a great expanse of golden hay. The smells were poignant too. Golden rod, purple thistle, black-eyed Susans, asters and many other flowers filled the air with their fragrance and yellow, eye-smarting pollen.

There were creatures, as well, that were so fascinating for me to watch. The prairie dogs were now in a frenzy to harvest every last weed seed they could find for they sensed the coming of winter.

The chill of the "northers" and their icy winds would soon be on hand. Winter brought forth the omnipresent snow, which drifted into vast, but irregular patches, forming checkerboards of dark windswept ground and glistening white drifts. Stranded cattle would often hunker down with their backs to the wind and press along the drift fences.

But it was late summer when the Plains were at their finest hour. The wind would blow as it always did over those high plateaus and the grasses would heel over, like a flotilla of sailing ships. In swelling wave-like motion, gust after gust followed to finally spend themselves at the edge of infinity. Colors would reflect like muted bands of neon as the textures of the grasses varied with the will of the wind.

In this rolling sea of grass I would often search for that fleetest of all North American mammals, the pronghorn antelope. Carefully I would sneak up the swales. Keeping my head down, I would stay to the lee side so as not to reveal myself by scent. I'd creep over the crest of the swale and then watch with glee to think I had succeeded, Indian fashion, to come within fifty or sixty feet of those fleet animals. I would lie there in the grass and watch while they browsed. I even saw their youngsters nurse in that nervous and sporadic manner which is so endemic to pronghorn young.

Eventually I would be spotted. Instantly their white rumps would bristle and form a glistening rosette. This strange heliograph-like flag would then alarm other nearby antelope. Thus, a chain reaction, geared for wild flight, would occur throughout the whole herd.

Even though I did not move a muscle the bucks would snort and mill in little circles, and then, in a flurry of beating hooves the whole herd would gallop away across to the next crest and down another draw.

These adventures were reminiscent of the homestead days in the '70s and '80s when my Great Uncle came to this country from Smaland, Sweden. For in those days the buffalo and antelope were a common and vital part of the American scene. The Plains were austere and uncompromising. It took men like him, men with vision to settle and forsee a future. My Uncle's first house was built laboriously out of sod on the slope of a hill.

I remember the stories that had been handed down in our household of how the corners of his building had to be shored with log posts. This prevented the buffalo from pushing the house down by rubbing their backs against the exposed corners. It wasn't that the buffalo had a grudge. Their shaggy coats developed perpetual and irresistible itches.

My Great Uncle's life was a story of vision versus adversity. His success was froth with hardship. His last house was a two story Victorian.

Thirty years after leaving my boyhood home in Colorado, I returned with our son. I wanted him to search and walk where I had walked. But, of course, there were only fragments of those grasslands left. The Rockies still formed that everlasting bulwark of peaks, but at their foot the Plains had changed. Now airfields and highways crossed the very draws and swales I had hiked upon.

Still, I wanted to see the Plains again, and so we journeyed to the Dakotas. Here were grasslands still left where the wind nudged the verdure with incessant waves as it had for timeless ages.

But there were also great silos and grain elevators and plowed fields with grain growing to the horizon as far as the eye could see. This was the vision of my Great Uncle. The grasslands had been his, yet he had not overgrazed nor over cultivated. He kept the land in a viable trust. When he was gone the land was as good as when he first broke the sod.

I think he had found the meaning of the Plains Infinity.

Pronghorn, Montana

Buffalo, Montana

Blacktailed Prairie Dog, South Dakota

High Plains at the base of Pike's Peak, Colorado

Badger, South Dakota

Red Fox, Montana

Buffalo Family, Montana

Spring on the high plains, Colorado

The Forestland Mosaic

It is not only the giant trees that compose the forest, but a look underfoot reveals an intimate composite of life.

A Vignette of Fall, Arizona

I have never forgotten the first time I hiked along the Hoh River and entered the Pacific Rain Forest. It seemed like the Land of Oz, except that I followed a chartreuse green path, for everywhere garlands of moss covered the landscape.

Filaments of moss hung in multitudinous layers. It crept over fallen logs. It covered rocks. It upholstered tree trunks. It obscured vision. The forest was an endless drapery of green.

The delicate chartreuse covering extends everywhere. No life form escapes for long, its insidious but beautiful applique. For a photographer it is a heaven, for here one can select a dozen totally different compositions by just turning the tripod head. Two steps beyond and another half dozen or more images seem to vie for selection.

While picture selection is infinite, ironically picture possibilities are slimmer. The photographer must finally come to grips with the limits of low light and the consequently slow exposure time.

And then the rain. It starts slowly, a fine mist slanting into the forest. It filters through the evergreens above and falls in a quiet, unobtrusive, percolating sort of way, diligently, continually, persistently. Drip, drip, drip those refractive, limpid beads incessantly gather upon the leaves and tremulously roll downward. And the small branches, weighted in turn by their new mobile burden, gently arch downward and form into drooping garlands of shiny green.

I strain to see beyond the foliage's lingering glisten. Where distant plants and trees stood, only vague shapes are visible. The forest has melted into a solid gray-green tableau, delicate, weird, beautiful, foreboding. I hunch my shoulders, check my camera for wet and damp, and press them close to me under my poncho. I trudge through the endless, dripping tunnels of leaves.

But one cannot wait for the rain to stop as this is the nature of the rain forest. Photography must be carried on even in the downpour. The best lighting conditions exist between the lighter rains for then the clouds do not hang too heavily. A picture session may only last twenty to thirty minutes, and then the downpour starts again.

While other forests may not be so wet the photographic technique is about the same. It is disastrous to wait for strong sunlight. Harsh lighting will always obscure and destroy the delicate forms in the shadows.

The aspen forests of the Southwest provide Romanesque arrangement of form. The slim, graceful white column-like trunks form impressive phalanxes. Grove after grove repeats this seemingly linear adulation to the heavens. And when fall and frost touches the leaves, the trees become a woodland spectacular of golden spires.

Mixed with the aspen and at lower altitudes are stands of ponderosa pines. In these expanses of conifers the Kaibab squirrel makes his home. This elfin-like creature is found nowhere else in the world except at the North Rim in the Kaibab Forest of the Grand Canyon. It is a medium-sized squirrel, but its fascinating and identifying feature is its all white tail and tasseled ears. He is a rare commodity even in his own habitat.

My first sight of the redwoods was flabbergasting. I had mentally envisioned just how big they might be by estimating distances. But to actually see them in their entirety was an experience I shall never forget. I could scarcely grasp their magnitude, and how was I to photograph those giants?

It is a challenge that I am not sure any photographer can amply fulfill. The towering, needled crowns provided a lattice so high that one could scarcely follow those stately columns to their apex. As skylight filtered through the lofty filigree it turned into a gray-green cast, illuminating the forest in a wan, nether world of translucent shapes.

In contrast to the giants there are the multitudes of small and tiny things that really make up the sylvan mosaic we call the forest. Fungi creeping along the fallen logs, blue and red berries hanging in the bushes, supple, lace-like, horsetails, embroidering the ground with enticing vignettes.

Underfoot, mats of fallen spruce needles form delicate patterns upon the emerald piles of club moss, beds of oxalis form green pillows of shamrock-shaped leaves. Flowers are everywhere. Dainty violets seem out of place against showy dandelions, the latter conspicuous both in bloom and seed. Each delicate white filament stands out in relief against the dark green of the background.

A freshet plunging between the spruce roots creates a continual murmur. The water spills over the moss-covered rocks and rushes onward. In the quiet pools water skippers float lightly upon the surface. Suddenly a water ouzel flits into the pool. The skippers, so numerous a moment before, are now hiding frantically from this hungry, feathered ogre.

In a crossbeam of light a tiny spider completes a round web between branches of blackberry. Its silken threads glimmer brightly in the sun. I bend down lower to examine it. It seems to come alive in a glistening, delicate weave of lace.

Often overlooked are the exquisite compositions created by the ferns that form a veritable forest of their own. Their graceful and brocaded textures, droop and spread over fallen logs. Mushrooms poke up from the duff floor, pushing their way with incredible strength. Some of the most common appearing can be a chef's delight, while the most colorful and beautiful can be deadly poisonous. The woodland ecosystem is revealed in a strange variance of large and small, exquisite and bizarre.

Maple Leaves, Arizona

Wild Turkey, Arizona

Winter Aspen, Arizona

Whitetailed Deer, New Mexico

Forest Evening, California

Beaver Dam, Colorado

Beaver, Colorado

Aspen Forest, Arizona

AF
VIII 72
R

Forest Fungi, Arizona

Redwoods, California

Kaibab Squirrel, Arizona

Spruce Needles, Arizona

Moss covered log, Washington

Fall Aspen Glade, Arizona

Olympic Rain Forest, Washington

Forest in Evening Mist, Arizona

The Jungle Enigma

Secretive plants and animals live mysterious lives under leafy, luxuriant canopies, sometimes 200 feet high.

Jungle Canopy, San Blas, Mexico

It was one of those wonderful mornings of joyful brilliance that flooded the lush growth of the tropics. Even though it was only 6:30 the sun had already drenched us. Everything seemed to have that sense of expectancy, a mood that I have come to know only in these latitudes.

We rolled up our cots, locked all the remainder of our valuables in the Carryall and waited on the edge of the tiny air strip. In the distance barefoot boys chased a herd of wayward Brahman cows off the runway to that the Cessna plane would have clear take-off. The pilot shouted in Spanish and waved his arms, then ran to the edge of the strip and shook his fist. As the assortment of boys and cattle melted into the brush we slipped into the cabin. We were so tightly packed that once seated we could scarcely move.

My wife held our eldest boy while the guide cradled our youngest son and wicker basket of cans. With cameras dangling about my neck I crouched in front with the pilot. In the baggage compartment we stowed the view camera and tripod, along with hammocks and mosquito netting, a batch of dry groceries, water melons and some live chickens and an iguana. Fresh food had to be alive. There were no refrigerators in the back country.

A jungle bush pilot doesn't waste any time. He had the engine almost wide open now and listened intently to its throb and resonance. Suddenly he released the brakes and in 50 feet our tails was off the ground. We were in the air after an incredibly short takeoff.

Of course, there is method to this madness, extra power means safety. Practically all jungle strips are short simply because it is such an extreme effort to hack them out of the forest. Maintenance is constant. Every three of four days someone has to go over the whole strip with a machete, not only to cut down the vining plants but to make sure that the thorn bushes are hacked out.

The air was like glass as he begun to follow the Usumacinta River. Even at that, I could tell the pilot had a distinct urge to get where we were going fast. Out of nowhere, clouds appeared. First tenuous wisps slipped by our wings. Then we went through more solid ones and by the time we were well on our way we hit turbulence from the steaming jungle below. "Muy mala," the pilot muttered.

We flew through a pass where it seemed we could easily have touched the tops of the great trees. They seemed to leap at us, literally in wave after wave until it seemed we were actually rolling over the surface of a vast green sea. Even our guide seemed to turn green. Suddenly we ducked into another solid bank of cloud which hung heavily between the peaks. When we popped out, the mountains were in back of us. We let down along the River on another tiny strip adjacent to the towering jungle.

The pilot shouted, jabbed his arms and hands at the rolling clouds and fidgeted with the controls. Then he waved us away. The clouds were now black and ominous. Jungle flying is safe only in the early morning hours. By 10 or 11 air currents and turbulence can make a death sentence out of a beautiful morning.

Suddenly he swung the plane around and was gone. Alone in the Central American jungle, we gathered our supply baskets, chickens, cameras, iguana, watermelons, hammocks and netting and trudged through the tangle of vegetation to our camp. We were at Yaxchilan, one of the most inaccessible of the Mayan Ruins. On our right across the river lay Guatemala and on our left Yucatan.

We had come to see the exquisite ruins, but I also noted with equal interest the fantastic array of plant life. Plants lay upon each other, grew upon each other, and died upon each other. Everywhere the design of the leaves were different, so many species were there.

The very size of the jungle canopy defies description. The giant sapodilla trees with their great flaring and buttressal trunks reach skyward 200 or more feet. Interspersed are a dozen more giants including the balsa and mahogany. All of these woody behemoths struggle upward in a never ceasing battle to survive.

After we had gone beyond the forest's outer wall the jungle assumed a different character and opened up. This was one of my greatest surprises. It was more park-like with giant trees spaced more or less evenly. Each had a tangle of lianas creeping up their host and in some cases the lianas were completely wound around dead and rotting logs. It seemed incredible that such huge, dead trees did not fall, however, they were completely and rigidly held in place.

The great lianas were everywhere, clambering upward, reaching, probing, always grasping, yet paradoxically, at the same time pulling, tearing at their lofty hosts with tendril-like holds. The base of these vines, trunks, I should say, were as big in circumference as a small city elm.

Other vine-like growths were just as incredible to behold. In suspension bridge manner the vines linked tree to tree. One of these was the water vine, which when cut can furnish gallons of pure water. Another plant of similar appearing growth was poisonous. Our guide used a vine with greenish bark for tea-like brews, which were quite refreshing. Colorful epiphytes and air plants clung to the trunks. Delicate lilies grew in swampy depressions. Strangest of all were the insectivorous plants one of which had a flower almost three feet across. Its death cup could trap a tiny bird or rodent.

One of the jungle sensations I became aware of was the smell. It was pungent, yet clean, for the forces of life conspire quickly to reduce the dying. In a short length of time fungus and the insects reduced the giants to powder.

The height, complete and absolute engulfment for anybody or anything living in the jungle made it impossible to get a bearing. No wonder the jungle remains such an enigma to man.

Jungle Floor, Yucatan

Jungle Lily, Guatemala

Armadillo, Yucatan

Evening in Jungle, San Blas, Mexico

Jungle Falls, San Andres, Tuxtla, Mexico

Air Plant, Yucatan

Kinkajou, Guatemala

Giant Saba Tree, Yucatan

The Swampland Mystique

The secluded swamps, bayous and other watery recesses are some of nature's most intriguing byways.

Lettuce Lake, Corkscrew Swamp, Florida

Long before I visited the swamplands I had a desire to see those strange, mysterious alcoves of hanging tillandsia, giant cypress trees with their knobby knees and waters lurking with gators. My concern was how would I go about showing that swamp mystique. How would I reduce the Okefenokee, the Florida Glades, the quiet bayous and those other places of quaking hammocks to 4 x 5 camera imagery?

It was a different camera approach for an old desert rat. I often examine a new area by just looking for its essence, its spirit, before I frame images in the camera. First, I had to define this new eerie essence. I saw at a glance that photography would require delicate subjective lighting to describe all those dark, obscure forms.

I considered this new challenge as I hiked into the swamp at Corkscrew Sanctuary. It was blazing hot and humid, and I had chosen summer, an unlikely time of the year to make my camera debut. I shouldered my equipment and with my son bringing the Hasselblad, 4x5 holders and our lunch we entered the swamp.

We were not prepared for the coolness. Layer upon layer of foliage above us was absorbing the sun's rays to an astonishing degree. It was a mysterious feeling. That strange unexpected calmness and solemn stillness seemed to brush our faces ever so gently. We were in a hidden realm of solitude and secretive life.

Soon my son beckoned to me. An eye had slowly risen up from the lavish covering of water lettuce. I would not have seen it had it not been pointed out. Camouflaged behind the plants, the dark body appeared to be a part of the swamp itself.

The eye rose higher. Then two, sinister elliptical pupils stared at us, surely sizing us up in a context of dinner. A long snout embroidered with dozens of teeth led forward from the eyes. A large gator, it moved silently through the floating plants. Strangely, scarcely a ripple occurred in its ghost-like wake. Solemnly, resolutely, menacingly it swam by us.

Suddenly an earth-rendering bellow and roar blasted away to our rear. I nearly fell into the water. The foliage and plants shook from the tremendous thrashing and splashing as the bull alligators fought. Though only thirty to forty feet away, we could not see them because of the green tapestry. Then, just as suddenly, silence. Silence.

I thought to myself, how could anybody traverse, much less survive in the swamp without knowing its secrets?

With gators and moccasins so much a part of the scene, we caught on to safety precautions fast. While in or near the water, we always stood back to back. My son scrutinized the water and surrounding area to our rear, while I concentrated on the camera. There in only a narrow angle of view when you focus under the dark cloth. Being gator bait did not appeal to us.

It was heartbreaking not to be able to photograph all the visual imagery. The light level was quite often too low. Even by tilting the film plane of the view camera, I could not get all the composition sharp. I either had to crop or be satisfied with a narrow depth of field. What photography I could do meant long exposures with big openings. It always necessitated very careful manipulation of the camera controls.

In the outlying islands and Keys I photographed the roseate spoonbill. A beautiful bird with white and rose plumage, but with an incongruous, out-of-proportion, spoon-shaped bill. Dead mangrove branches were their favorite perches. Pink forms reflected beautifully in the brackish water inundating the mud flats. Out on the Keys we found the tiny, Key deer. The bucks were so small they seemed like playthings.

The hardwood swamps of Louisiana provided me with some exquisite imagery. Maples with gray trunks, splotched with light and greenish lichens grew in picturesque stands along the edge and in the slow moving, bayou water. Many of their leaves had fallen, and as daylight filtered down through the trunks, glints of amber sparkled back from the brown, wet, floating leaves.

The Glades were unique. They are fed by an overflow ninety miles wide but sometimes only six inches deep emanating from Lake Okeechobee. This strange flow of water nurtures the Glades and creates probably the largest subtropical swamp in the United States. The Glades are so flat that the landscape reminds you of the Plains. Only an occasional rise occurs here and there. But these "hammocks" are covered with an astonishing array of plants, all of which in turn provide a secluded haven for wildlife.

Most of the hammocks revealed luxuriant growths of palmetto. Their fan-like leaves offered exquisite compositions, particularly when the rain had garnished the leaves. The gaunt, bald cypress often seen growing in clumps and groves towered more than a hundred feet above the hammocks. Between these elevated masses of vegetation grow continuous expanses of sawgrass. Scarcely a grass in the normal sense, these hardy blades have wicked, ragged edges. It is difficult if not well-nigh impossible to walk through this grass. In a matter of minutes one's clothing can be cut to shreds.

One of the most poignant, impressionistic scenes of the South was the Suwannee River. Luck was with me since the sun had been playing hide and seek all day. Patiently I waited for the right intensity of light. Then, slowly it brightened on the Suwannee. The River's stately flow shone in a progression of quiescent ripples each picking up a new moving catchlight. In the foreground, palmettos and cypress framed the sparkling surface. A tapestry of muted green hung in the background. In a moment the nostalgic mood of Stephen Foster's "Suwannee River" materialized.

Photographing that lazy stream seemed to conjure up all of the patina and mystique of the Southern swampland.

Corkscrew Swamp, Florida

Alligator, Florida

Louisiana Heron, Louisiana

Hardwood Swamp, Louisiana

Roseate Spoonbill, Florida

Great White Heron, Florida

Everglades, Florida

Okefenokee Swamp, Georgia

Key Deer, Big Pine Key, Florida

Palmetto Thicket, Georgia

Duckweed Swamp, Florida

Cypress Swamp, Louisiana

A Personal View

A photographer's critique provides insight in the exacting art of nature photography.

Tundra and Dall Ram, Alaska

The Beginning

I have always felt that there are certain times that, in retrospect, one realizes are life's pivotal points. Certainly my fourteenth birthday was such an occasion.

My mother had given me a Brownie box camera.

As I look back at the event, I think the gift had a two fold goal. It was intended first to provide me with a new experience; and secondly, hopefully, to rid our backyard of my zoo, for trapping and bringing home small animals was an obsession with me. Her reasoning: ''He would take pictures and forget about his fixation with Frank Buck and 'Bring 'em Back Alive'.''

Along with other explorers of the day—Adm. Byrd, Osa and Martin Johnson, Father Hubbard, etc.—Frank Buck went places and did things. In remote jungles he trapped and brought back fabulous, live birds and animals. I'd often lie awake at night thinking about those amazing globe trotters and their exotic adventures, and how I could emulate their fascinating lives. In my imagination I seemed to be only a step behind, once I had my camera.

So my mother's reasoning backfired. My collection became dearer to me yet because I used the camera on my animal subjects. Unfortunately, few of my shots were any good. With a wire mesh screen in the foreground, the caged rabbits, skunks, squirrels, etc., always appeared to have that furtive, jail-bird look. It didn't take me long, however, to start experimenting. I tethered my unruly animals and then shot pictures. I even tied, with special and diligent care, a couple of my skunks. I'm sure the latter was devotion beyond the call of duty as I was always fearful of retaliation. It couldn't be compared to a lion chasing you across the African veldt, but it did take courage of a kind. My photographs were better, but I couldn't stand to see my precious animals leap and tug at their thongs. The solution lay in going to the animal's habitat.

Besides, I needed to see different places so I could use my new camera more. I made mental inventories of where I could find my quarry. Plenty of wild places lay close to where we lived. In that fourteenth year, my odyssey took me over much of the countryside near Colorado Springs. I hiked and biked to Fountain Creek, Prospect Lake, the plains in the back of the Printer's home, Austin Bluffs and the old cemetery.

I was both Tom Sawyer and Huckleberry Finn in those sublime days. But instead of the churning Mississippi being the common denominator of life, I always had the pristine Rampart Range to focus my thoughts upon, for there towered Pike's Peak, that colossus of nature. Usually covered with snow and soaring into the heavens, it was stupendous to behold. Gazing at those forested bulwarks of tilted granite I always felt at peace. Like a great crenelated wall, Cheyenne Mt., Mt. Rosa, Mt. Almagree, Cameron's Cone, Pike's Peak, Mt. Manitou and a myriad of other heights chained northward. That extravagant wilderness always reached out to me, completing a magical synapse between nature and boy. I would often lay back against a rock, content just to study the clouds that would suddenly well up in pile upon pile of silver-rimmed gray over those sparkling peaks.

For a boy imbued with the thrill of seeing his very own pantheon of nature, I occupied a unique area and place in time. I had my heroes and everywhere I was surrounded by nature's bounty, a bounty that beckoned to me in continual attraction.

My collection of animals began to taper off. Now I stalked my quarry carefully and sat unobtrusively near their burrows and dens so I could snap them as they scooted out. Most of my camera results were mediocre. Usually, the image was just a blur and the forms unrecognizable. But I was learning and discovering a great deal about ecology and the creation of imagery. Even at that young age I was becoming a naturalist. To get my quarry into camera range I needed to know about animal behavior, and I learned that animals were individualists. The stalking technique that worked for one might not be applicable for another. I laid out crumb bait trails and built blinds. I imitated calls and cries. I watched and observed and studied. I also read a great deal more after my mother introduced me to Ernest Thompson Seaton, an early natural history writer.

Then I began to frequent Evergreen Cemetery just southeast of the city, not so much in reverence for the departed, but to search for wildlife that found haven and retreat among the stone walls, terraces and trees. In my obsession to photograph animals, I became one of the hunted as well. The salty and humorousless ground foreman of the cemetery always kept an eye peeled out for me. Brandishing tree clippers when he found me looking for animals in the old part of the graveyard, he kept me sprinting over the headstones to avoid a showdown. When he did corner me, his verbal lashing was severe.

Rock and grass, Arizona

There was no doubt about it. In my small world, the cemetery was where the action lay. On one of my cemetery trips I chanced upon a grave plot of Helen Hunt Jackson, whose poetry and writing sentiments, like that of Emerson, Thoreau, and Seaton, were the forerunners of enviromental concern. The poetess had special significance for me, since we had often recited her lines in the lower grades. In fact, the grammar school I had attended was named in her honor. Her verbal imagery had always excited my imagination and now, in a way, I had a personal introduction. There in back of the stone coping she lay. Somehow her presence brought a subtle change in my outlook. I began to sense a relationship between mental images and visual images. I often would ask myself how her phrases and verses would appear if I could actually see their meaning in a camera. I was beginning to understand how the camera might communicate a visual language. I would often examine other old grave plots. Headstones of Civil War veterans always intrigued me with their clipped, terse sentiments chiseled into the gray Colorado granite. I think those messages posing such solemn moods sharpened my mental imagery. I was continually imagining how these people would have looked, had I actually seen them.

My first attempt at photographing a story occurred when I decided I must take a sequence of pictures of young magpies in their nest. They are picturesque birds, slightly smaller than a crow. Glossy black plumage is accented by a white patch on their wings and chest. Their tails are long, supple and jet black. Their neat, tuxedo-clad appearance is in dramatic contrast to their hoarse and raucous calls. First, I had to scale a tall pine, and then I had to judge the right angle for light to enter the nest. The latter was not an easy task since magpies build a heavy stick nest complete with a sturdy roof. Finally, I got the entrance enlarged enough to make pictures, but by then, it seemed the whole magpie citizenry of El Paso County, Colorado, had convened to harass me. I could not think, much less perform with my fixed-lens box camera. In a flurry of whirring wings and stabbing beaks I called it a draw and slid down the tree.

The first really clear picture I took (in those days we always called good pictures ''clear'') was of two horses on a slope. Long grass waved in the wind, box elder trees framed one side of the sloping hill, the Rampart range stood stark and foreboding to the west. What a thrill overcame me when I finally pushed the shutter lever. When it came back from the drugstore I was elated. The photograph had a genuine composition to it. It seemed to say more than was actually represented in the print.

Film processing was cheap in those days, but I was still spending more money than my mother could afford. Finally, I got the word. Conserve. The alternative was to process my own film. In the '30s it was a simple matter; you just went to the main drugstore in town and bought a tube of MQ developer, seven cents, and a little packet of fixer, 15 cents, called hypo. Fifty small sheets of contact paper cost ten cents.

I processed at night on the kitchen drainboard. For tanks I used breakfast cereal bowls. Of course I didn't have reels. I would just dip one end of the film in the developer, then slowly pull up that end as the other end was submerged. I fashioned a contact printer out of a fruit crate and old Christmas light string. When my first print came up I was so thrilled that I yelled! It was a miracle!

In those days a roll of 620 film cost 23 cents. Even at that, when I walked to Colorado Springs to buy my meager supplies, I was often tempted to stroll by the Chief Theatre and look at the marquee. There in gun-toting glory the cowboy show of the week beckoned, featuring either Tom Mix, Hoot Gibson, Gene Autry or Tim Holt, favorite film cowboys of the day. With a movie costing a dime, a youngster had a lot to rationalize.

I became more proficient. With a bit of money saved I bought an Agfa-chief. This was also a 620 camera, but gloriously, it had a lens that could be screwed in and out for close or far focus. It also had a little circle of yellow cellophane attached to the end of a lever which I could use as a filter. And, best of all, it had an eye level viewfinder. Now I could really follow action. By this time I was in high school. I had won a small award. One of my pictures was standing in the window of a photoshop on Tejon Street in downtown Colorado Springs. I also took a few pictures for the high school annual, *The Terror Trail*.

With more demanding work, I discovered I had to have an enlarger. Buying something of that magnitude was out of the question. The alternative was to make one. A five gallon syrup can became the light container. I bought an old Recomar, 6cm x 9cm size. Lens, focusing track and bellows were bolted to the bottom of the can. I fashioned a slideway by cutting out part of the camera's back. For a carrier, I hinged two pieces of double strength glass with bandage

Mountain Goat, Montana

tape. A blacksmith constructed a waterpipe standard, which I bolted to an old breadboard. The can moved up and down by the use of a pipe sleeve. It worked fairly well considering I had less than four dollars invested. For large printing trays I used Mom's enameled baking pans.

It is surprising how inventive one can be when lack of money sets the pace. I've built telephoto lenses by using cardboard tubes, car cylinder sleeves or aluminum tubing. I've fashioned micro switches from thin brass sheets and hairpins. When I needed filters I made them. Actually, it was easy. I put food coloring in hot clear Knox gelatin, stirred well, then poured a little of the solution onto one of my mother's fine china plates. The liquid soon solidified, then dried, colored, but ever so transparent. Because of the china's glossy, non-porous surface the dried gelatin peeled away in a uniform sheet. Then I simply cut a square or circle the size of the lens mount, used airplane cement and a cardboard strip to reinforce the edges. With care such a filter would last several weeks. I am sure I could never have engineered my ideas had it not been for my continual experimenting and the fabrication of unique devices. Lack of money, though certainly inconvenient, scarcely ever stopped me. Different ways always seemed to present themselves, and then with new inspiration, I would redesign the whole project.

After I graduated from high school, we moved to Arizona. My photo experiences began to pay off. I worked more on yearbooks, and made publicity pictures for Phoenix College, and later for Arizona State University. It all helped to meet my college expenses. I also received a scholarship due to my photo abilities. Slowly but surely, event after event dictated that I continue with photography as a profession. Upon graduation from Arizona State University, I began as a staff photographer on the Arizona *Republic* and later joined their Sunday supplement magazine.

During this time I met a sparkling young lady who worked in the national advertising department of the newspaper. We squeezed in a marriage between my assignments, and later our three children were born, one daughter and two sons. Many times we have worked together while on photographic trips. In the early days the kids packed the lunches. As the boys became older, they shouldered more of the camera equipment, which always did and still does include the 4x5 view camera.

Of those years as a news photographer I could write a special book. Events, people, politicians, in fact, my photography coverage of the Arizona scene, along with other staffers during the '50s and '60s was a visual fabric, a history of a burgeoning Arizona. There is no other job quite like newspapering. Working out of a city room of a large metropolitan newspaper is a college education in itself. I can scarcely begin to name all the important people I've met, listened to, lunched with, interviewed and photographed. The resulting news features and the news photography are indicative of how important the media can be to our lives.

I had mixed feelings about leaving the *Republic*. But my desire to specialize in nature and environmental photography had become a commitment. I had had a great education on the nature of man, but now it was time to move on. I had to explore in depth that mystical, yet real world of natural wonderment. I also had another reason for leaving. It was a future desire to teach. With a few added years of free lance work I felt I would have a great deal to say and offer, a philosophy stemming from all this professional experience. I would, by sound example and disciplined skill, share my enthusiasm for photography with young people who sought it.

Free Lancing

I don't think any photographer can begin free lancing at an instant's notice. It takes a great deal of time to build up a clientele. That's how it was with me. During the time I was with the newspaper, I made contacts and did photo work for magazines which were devoted to scenic and wildlife themes.

Whenever I could, I would try to meet editors and art directors and show them my newest portfolio of creative efforts. Strangely, I met many of the editors only through the mail, that is, by correspondence. Even after all these years, I still haven't had the pleasure of meeting some of these old friends in person.

Though competition is keener than it was years ago, I believe that the opportunities are greater. There are more magazines published now than ever before. Most are specialty magazines, many of which relate to the outdoor field. All must have good photography in order to exist. And book publishing has literally mushroomed. All of these publications need competent and sensitive photography.

Most free lance material goes through the mail. It must be carefully documented, and captioned accurately so that there is absolutely no question in the editor's mind about the story content. As soon as the editors get to know your style and to rely upon your integrity, the way is set.

One of my first nature stories concerned Arizona's beaver. These animals lead fascinating lives and appear to be so intelligent concerning their daily activities that it occured to me such a photo essay would have a great deal of readership. I spent days getting pictures. In fact, it took the better part of a year.

I would leave after my shift late Friday night so that I would be in the mountains by sun up. Carefully, I would place my Speed Graphics so that they would cover a selected portion of beaver dam, woodyard trails, lodge slides, or pond, etc. Sometimes I set the cameras so that the animals would trip the shutters simply by hitting a tightly drawn cord. When that did not work well I devised remote control solenoid switches to activate the shutters. Thus, I could watch through field glasses and control the composition. At night I used a high-powered flashlight. By watching, I always know which end of the animal faced the camera, obviously pretty important, but not always guaranteed when the animal took his own shot.

The big problem occurred when I had to change holders after each shot, which necessitated disrupting the animals' activities. When the beaver sensed my presence, ''kerplunk''would go their tails, crashing down upon the water with a terrific slap. Of course, this alarmed other beaver along the creek. In like manner, they also banged their tails, and so this strange alarm system sounded up and down the creek. After a bit of time, when all was calm, my subjects came back. Soon the beaver began to regard me as a part of their environment.

The completed story first appeared in *Arizona Highways Magazine,* and then much to my amazement, the *National Geographic* sent a telegram indicating their interest, asking would I be so kind as to write a different story and obtain a few more pictures for their magazine.

Yes, indeed. I certainly would!

Later *Sports Afield* wanted it. Then I received a commission from one of the children's encyclopedias for a similar story. It all goes to show that unique pictures can create their own market. In all, that series of beaver pictures sold more than a dozen times to various publications.

I discovered quickly that readers love animal pictures. I had also stumbled upon a new concept—life cycle portrayals. In the nature field my newspaper experience was paying off by applying this new dimension. I ''featured'' the animal as he lived in his environment, and described him as a personality, as well as his activities throughout the seasons.

My free-lance career jelled with a story on the kangaroo rat. Here was another natural. No magazine had published pictures of these animals, at least not good ones. I collected them and raised them. I studied them on the deserts and at night I photographed them. This nocturnal creature had adapted so well to desert life that it seemed incredible. To survive, the rat could actually manufacture water with his own metabolism.

Such a fascinating animal had to have its life portrayed. However, the connotation ''rat'' had to be overcome. I wanted to show this creature as a perky little fellow, intent on his own business, and making his way successfully in a very hostile world.

To show this spirit of brightness, I lowered the camera lens almost to the subject's eye—always a good way to present small animals. It gives small creatures stature, dramatically punctuating their stance. The angle becomes at once eye catching and different from a casual observation.

To make the theme of the night paramount, I double exposed the film by taking the full moon soon after it had risen with a 20-inch lens, and then exposed the same film again with a 5¼-inch lens. This put a large moon over the shoulder of the subject. The result; a pert, little guy, nonchalant, full of energy, emerging from his burrow in the moon glow, ready to take on the world. For lighting of the scene I used double electronic flash.

A month after my story appeared in *Arizona Highways Magazine,* an offer came from a book publisher to use the rat story as a children's book. In addition to that, the book's editor wanted three more animal stories which I had done for other publications.

The fact that I want to emphasize is that all of these stories had been handled in my new approach. The story was very personal and the pictures showed more than a bundle of fur. They portrayed a personality, a being, a spirit, a life that was unique and fitted into its very own special niche in the environment.

Any nature photographer, to be successful, must provide those insights into animal behavior for the viewer. He must also ask himself, ''Am I portraying the essence of the subject so that the viewer discovers the subtle nuance of life which I am translating?''

Photographing wildlife can lead to strange encounters. One of the most bizarre occurred when I was doing a story on the great horned owl. We were camping with our 18 month old daughter for the first time in the pinyon rim country of Central Arizona. I had located an owl roost a few days before. The adults had begun to accept us, perhaps still not as part of the scenery, but they tolerated us to the extent that I could set up the electronic flash equipment near their feeding tree.

I was determined to get flight shots that night by baiting them to a certain branch. Soon after dusk I began ''calling them in,'' that is, imitating their hoots and cries. It wasn't long before one flew low over our heads. In another minute a second owl swooped low over the branch and then our camp. My strategy seemed to be developing nicely.

Then suddenly the first owl landed on the rail of the portable crib where our little daughter slept. We froze in our tracks. Like all hungry raptors, hawks or owls, they will strike at anything small that moves. When you consider that the adult birds were equipped with talons—eight, two-inch, curved scimitars apiece, you can understand our consternation.

I distracted the owl's fascination with the form lying in the crib, only to have a second, and then a third owl swoop down from our rear. By then my wife had thrown a jacket over the baby and was frantically stuffing her into a sleeping bag. Being a long ways from the Carryall, we stood guard the rest of the night. That beautiful evening turned out to be a nightmare. And there wasn't much else we could do, for we often camped without a tent.

A strange sequel occurred a couple of years later, just after the story had been published. Two Indians from a reservation near Phoenix came to our door one night. They had concluded that my owl photographs published in *Arizona Highways Magazine* were linked to a death which had occurred in their clan. I explained the pictures were taken just because I happened to like birds, and I assured them that no such consequence could have happened. After several cups of coffee and handshaking all around, we parted friends. By then it was two or three in the morning. We were completely dumbfounded and exhausted. At least my photography had a following.

Buy the best equipment, certainly a most important consideration, the kind that will give you the best technical quality. It isn't much more expensive when you leave off the frills and gadgets. They add considerable cost and they also add weight. Remember, you create the image, the camera doesn't. When you get right down to the basics a camera can only provide three functions: shutter timing, focus and aperature light control.

I've always tended toward larger format. For all my animal and bird shots I scarcely ever use smaller than 2¼-inch square, though even at that, when you consider vertical or horizontal shots from 2¼-inch square there will always be some frame area loss due to cropping.

At times it gets downright tiresome to backback an *f*-3.5 300mm TEWE, or *f*-5 500mm TEWE, or *f*-5 600mm TEWE Votar, or an *f*-5 800mm TEWE Astragon or even two of these long lenses that will cover a 2¼-inch square format, but remember the editor isn't a bit concerned about how tired you may get. He wants quality: sharpness, correct exposure, striking, dramatic lighting and a stance of the subject that portrays a message.

To check the quality of a color transparency always use an eight power or more magnifier. If the image doesn't show edge sharpness it is too soft for reproduction. The most difficult thing in the world is to throw away those prize shots, but if they can't be used, what can you do?

For the large animals I usually use the *f*-5 600mm lens. The long focal length insures that the subject will be separated from other planes in the pictures. It will also foreshorten the perspective so that the subject stands out boldly.

Hand held images just do not have that crispness that sets apart a quality image from merely an average one. It is imperative to use a tripod with long lenses. I never use a camera gun stock. For one thing, they are not steady enough and secondly, you have to decide whether you want to skeet-shoot or make photographs. For rugged, outdoor work, wooden tripods are best. Salt water doesn't corrode them, sand and mud cannot damage them, and a bad fall will not dent them.

Wilderness photography usually means backpacking because it is simply out of the question to hand carry equipment. One doesn't have enough arms, and secondly, it is too tiring. I always use a pack that I can jettison immediately in case of a trail fall, rock slide, or even a bear encounter, which is always possible.

I use a long lens with a 1000*f* Hasselblad. The camera is a ''golden oldie,'' but for a long time it was just about the

Bristlecone Pine, Nevada

only camera available that had the shutter in the body. This enabled me to use a number of different focal length lenses. The only problem was to design a lens mount.

For special, larger format animal and bird telephoto pictures, I devised a viewfinder-rangefinder combination, which focuses with a Bausch and Lomb 508mm aerial lens, mounted on a 3¼x4¼ Speed Graphic. For general nature scenes, I use a 4x5 Graphic View camera. For quality results there simply isn't anything quite like a view camera. If need be, there are perspective controls which can add a great deal. The tilting film plane is a must for needle sharp focusing, a flexibility that cannot be minimized. It permits the photographer to control the focus from just inches in front of the lens to infinity.

The larger format all but eliminates grain for highest quality black and white prints. By being able to change a variety of lenses, the 4x5 becomes a real work horse. I use five lenses with this camera, a 90mm, 127mm, 165mm, 205mm and 300mm, depending upon the degree of control and coverage angle that I want to emphasize.

Naturally, all these odd camera lens combinations eliminate any sort of metering through the lens. I've always felt that a good hand meter is far superior to a coupled aperture-meter combination. Metering through the lens tends to make the photographer passive toward the light situation. He may cease to be creative and rely too much upon a mechanical system to compute exposure. Rather, he should evaluate the subjective quality that molds his subject matter, light.

A light meter is only a tool, and a tool that must be regarded with some skepticism. By that, I mean, readings must be interpreted for scene effect. The image may need a stop more or less than indicated. I take readings from the back of my hand much of the time, especially for far away scenes. I assume that the light on the distant object would be of the same reflectance as on my hand.

To really understand exposure you should process your own film, even color. Then you control all the variables. Eventually, you will become so attuned to your system that mistakes in exposure rarely happen. Technical expertise must become a natural reaction. All efforts then go into the making of the image.

More time should be devoted to abstracting bits of life from the whole environment. This is an important consideration. Invariably, inexperienced photographers try to include too much in the frame.

There is an Art of Seeing

The romantic appeal of nature photography has brought literally a legion of new photographers into the wilds, all within the last few years. When I started out upon the outdoor free lance trail, there were only a few professionals. We would often meet on location, Yellowstone, Alaska, Canada, wherever the season dictated action in that area.

We were a special breed of photographers, bent on making the very best images for our various publications. We would often drink coffee together in the evening around our camps and tell innumerable funny, sometimes harrowing, tales about our encounters with wild animals. There was solid shop talk, too, and we learned a lot from each other. But when we were on location, we were fiercely competitive, because it was our talent, our artistic skill and technical know-how that would pay the bill.

There are still the active pros in nature photography, but they are now far outnumbered by the sheer multitudes of camera fans involved in outdoor activities.

Regardless of the objectives, be it only a few hours' interlude in the park, to perhaps a lengthy photographic examination of the wilderness, the urge is there, perhaps even an ancient urge, to communicate with the pantheon of nature.

For a moment, all is forgotten. It is an adventure steeped in exhilaration. A bit of life is forever documented, along with our own confrontation of self. It is an instant bond between ourselves and subject, because the act of making the image is a very introspective and personal one.

All this creative effort is compounded when you consider that no two photographers are ever going to see the same image exactly in the same manner. Seeing an image is simply a matter of abstraction. The photographer should always ask himself, "Is this the simplest way to portray the subject?" Nine times out of ten, the simplest is the most effective, and certainly, simplicity is elegance when handled in a creative manner.

By repeated selection, by constant redefinition we photographers seek an intimate concern with the essence of life. Considered in its entirety, nature photography does, indeed, become a tremendous influencing aspect on the artistic taste of society.

A typical forest scene, for example, may have too many elements in the confines of the picture frame, perhaps too many trees, too many shadows, too many rocks, to be really effective as a communicative image.

The solution is easy.

"Select the most vital, the most vibrant portion of the picture arrangement," I always tell my students. "Thus, the whole imagery of forest may be reduced to a glistening droplet of dew clinging to a branch of pine needles, or a mat of fallen leaves lying between the roots of an old stump, or a clump of fungi creeping along a rotting log, or a beam of light breaking through a filigree of leaves."

Any one of these approaches may tell the story, or reveal the essence of a forest more poignantly than an overall view. The result is a selective and more powerful way of seeing. The same procedure is true in photographing other ramifications of nature.

There are times, of course, when you simply cannot reduce your image to such a small vignette. Then your pictorial is of such vibrant strength and outstanding composition that such a procedure would water it down. The essence, then, becomes the whole scene and this is what you want to communicate.

Every nature portrayal or composition, as in any creative expression, must have a theme, or a statement, with which the viewer can identify.

Before shooting, I feel there are a few mental aspects which a photographer should consider. They are compositional exercises, or more simply stated, a means for employing more emphatic ways of seeing.

Visual perception is the key. For information we rely more on our eyes than any other sensory organ. Using our eyes we constantly probe, dissect and evaluate. The more we seek in visual meaning the better we transfer this process into a coherent, pictorial statement.

Consider that all plants and wild creatures are living in a mosaic of mood, color, form and design. And this composite interpretation of subject and environment is what we seek to relate in the image and to communicate through a composition.

Put another way, I dissect and rearrange the configuration of the composition to determine other viewpoints. I study them. I decide what I like best. In this way I discover how the magical set of elements combine to create the unity of the photograph—the design of the image or, in short, the composition.

Composition is an arrangement of space, a structure, a group of design elements that relate to one another. Thus, effective communication in a photograph demands that one must work with the design components at hand rather than follow preconceived rules, which may curtail any expanded way of seeing.

If you need to change emphasis in a picture structure, then the art of the matter is to change the angle of the camera to foreshorten the perspective. Or perhaps it would be best to move closer, or perhaps a portion should be deleted, or simply add an area by backing off with the camera.

I use the viewfinder as a masking device to assess the composition. I probe the scene. I look for the limits of the composition, then I focus for needle sharpness if the image demands such treatment. It is an exercise in discretionary viewing. It seems to me that many photographers never really understand this point.

Spacing of items is important. The picture should be composed so that the result is balanced. Dividing the proportion of space within the frame's format is best examined and experimented with by again moving and placing the camera at different angles.

For example, I balance one large object with smaller objects or shapes. A sense of dominance, or even tranquility can be achieved by balancing one picture area against another.

Using repetitive patterns is a good way to insure a sense of rhythm. Conflict in picture emphasis vanishes. Consequently, the effect is pleasing. There is a powerful sense of order to be derived from a repeated design.

I try to determine how perspective will influence the image. Linear and angular perspective can create a feeling of depth, an exploration of distance, a search for infinity.

Another method is to emphasize relative sizes. Consequently, the subjective mood of height and weight is easily felt, as well as space.

I take advantage of atmospheric disturbances such as haze, fog, dust and clouds when I can to indicate picture depth. Showing the difference in tonal values between receding planes in panoramic views produces a tremendous mood of distance and gives the composition a feeling of life.

This mood is particularly prevalent in a shot I took when my son and I were camping in the Red Rock country of northern Arizona.

Goosenecks of the Colorado, Utah

We were hiking to Spider Rock in Canyon de Chelly. It had been raining and I wanted to get to the overlook by sundown, since it seemed as though the weather might break. Just as we unpacked the view camera at the Canyon's rim, the sun came out.

My son gasped. ''Hey, Dad, you're a genius. How'd you know we'd have a rainbow?'' I smiled back. ''You've always got to be ready,'' I said.

In arranging the composition the photographer must be aware of negative space. Simply stated, it relates to shapes which surround the main subject. The main design and composition may not actually be the leaf, or the rock, or the animal, or whatever the photographer centers his viewfinder upon, but the area which surrounds the subject.

Composition, then, I should emphasize, is always a full and complete arrangement of space including all design elements with a photograph's format.

While certainly not inclusive, these few functional ways to see and arrange a photographic image will enhance anybody's creative efforts in photography.

I must point out again, these thoughts are not rules, but only methods to gain better insight into a way of seeing.

Light Structures the Form

After composing the image, the rest of the picture outcome will depend on how well one resolves the problem of controlling the light. Remember, we are continually working with a two-dimensional medium. It is the photographer who must manipulate the light to eventually structure the form and show picture depth.

It's remarkable how we take the sun for granted.

It is the regulator of all our lives, for every life function in our solar system depends upon its radiance. It is no wonder that our ancient forefathers paid such homage to it, and yet, in photography so many photographers, even professionals, seem to disregard its rays. They disregard the light in the sense that this radiance is not utilized with discretion and facility. They light a subject rather than structure it. Their imagery might otherwise be enhanced a thousand fold.

While the sun is the key light of creation, it is also the key light of all natural light photography. Its rays structure the forms that we see. It creates the highlights and the shadows. With the sun we can backlight, sidelight and frontlight. We can, in short, create an infinite number of ways to describe a subject.

An intriguing relationship exists between the compositional design and the lighting angle. They must complement each other and if they do not, then it becomes the task of the photographer to see that they do.

One dramatic way to achieve a third dimensional feeling in a photograph is to overlap the highlight area against another shadow area. For example: should you be making a pictorial of two trees, it might be an excellent idea to hold the camera in a position, where through the viewfinder, the highlight of one trunk is seen to overlap upon the shadow area of the second tree. Notice how this vantage point gives you a third dimensional view. It will indicate picture depth (not depth of field) in the photograph. The composition is strengthened. The differences in light values now produce structural forms and receding planes and distance.

The time of day, too, dictates the lighting conditions. With the sun directly overhead there is little modeling, and despite the advice given by the manufacturers of film, this span of time, during the middle of the day, is usually not the best time for nature photography.

When the light is low it provides a better modeling source. Then, with the sun to the side, examine the image. See how much better the typical outdoor photograph becomes with the use of cross lighting.

The same lighting technique is used for photographing textured surfaces, one of the most fascinating prospects of nature photography. Every surface—sand, wood, water, etc.—has a texture. Shiny, slick surfaces are textures as well, but may be more difficult to show and define.

The challenge, then, is to indicate texture in a dimensional manner. The answer is in applying some sort of backlight, or cross lighting compositional approach. Late afternoon, or early morning, is usually best. It is then when long shadows occur, and thus, the surfaces with their irregularities will be more prominent.

As the sun sinks lower, even fine objects can be made dimensional. Watch through the viewer. Compose the scene by looking toward the sun or slightly away from it. Every little pebble, rock, root, driftwood, dune, tussock, tree, or distant mountain will be dramatically accentuated. Now, turn completely around with your back to the sun. Immediately you will discover that all the dramatic qualities of compositional pattern, design and form have vanished. It is your discretionary manipulation of the sun's radiance that has made the picture, dramatized the composition, then enhanced the subjective third dimension.

The designing manifestation of sunlight has shown the structure of the composition. By backlighting, sidelighting or lighting in some given angle, the modeling produced actually becomes the skeleton of the composition.

Another important factor concerning light is its quality, its intensity, and its color, too. If it is direct, many of the circumstances already discussed will apply. But if the sun is diffused by overcast, or if the light is completely incidental with no directional force, it can change the overall structure effect of the composition drastically.

There are times when an overcast sky is needed. Soft lighting will eliminate all definite shadows. This is best for thick foliage, dense forest and the like. Excessive contrast cannot produce intimate details. In such cases, subtle tonal values are considerably more important. The gradual change in tones themselves constitute design and structure. Dramatic, contrasting highlights would defeat the subtle composition of the image.

Contrast can always be a problem. There is a time for it, and there is a time when it may be detrimental to the composition. There is a technical problem as well. The lower light values may necessitate extremely long exposure.

The evidence again suggests that the photograph should not be governed by rules, but by observations that best identify and demonstrate an understanding of subject matter.

Light never ceases to change. The photographer attuned to this state of flux can find solace if not inspiration in these challenges. Light is such an integral part of photography that we tend to overlook its complexity.

It may be simplistic to say that control is the answer. But it does become apparent that by examination and manipulation (by manipulation I mean selecting the time of day, the angle, the intensity, the color, etc.) of light will eventually separate the photographer who "hopes and snaps" from one who evaluates and produces.

On Location

Nature photography is an alchemy of all the intrinsic art values so far discussed. Such prerequisites may seem formidable to a beginner; yet it does not take long to acquire a bit of sophistication in image selection.

Droplets of dew glistening in the backlighted brilliance are always eye catching. Sparkling water upon leaves and grass can be a real challenge. Ferns, leaves and flowers with their variant shapes and vibrant colors are exciting to behold, particularly when you see the arrangement in your viewfinder. Textures in stones, water, beaches, cliffs, sand, pebbles, wood and roots have their own aura. Try producing a photographic statement of the subject's content and feeling.

Ferns can provide a rhythmic pattern, and are wonderfully pliable, that is, they can be arranged and rearranged time after time by simply changing the camera's angle. While all these scenes are comparatively easy to photograph, always remember that you control the composition. When you choose the time of day consider the lighting. The camera angle will control the perspective, or how you see the subject.

The way you envision the composition will become your style, a dramatic photographic approach of your very own. It will become your signature. I always like to emphasize this to my students. This is why it is so important to find one's self photographically. There can be a tendency to become a follower of a movement, to seek a brand of imagery which may sweep you along with it and a veritable multitude of other photographers. It can be tragic to discover that your photography may be similar in compositional style to the rest of the group. Precious time has been consumed in a virtual copying session rather than following your own inspiration. Always experiment.

But start with simple subject matter. Whenever your subject seems to evoke an emotional response, and whenever the viewer receives an emotional jingle, you, the photographer, have communicated your statement well. This is the feedback for which all photographers are trying.

Small wildlife present a challenge to the beginning photographer. They are best attracted by food. Thus, it is far easier to bring animals to the camera where sunlight is available, or where supplementary light can be set up.

Dandelions, Colorado

Ringtail Cat, Arizona

Most wildlife have definite foraging habits. In nine chances out of ten, small animals will frequent an area at the same time every day or two. Such regular habits are also displayed when they leave, or return to their nest or burrow. And, I might add, if it were not for this significant fact, nature photography would be considerably more difficult.

After studying such behavior patterns, it is a fairly simple matter to ''cut the trail'' with tasty and aromatic bait. Do this by laying a trail at right angles to the foraging area. Place a larger pile of food against a suitable photographic background. To decoy wildlife and induce reticent animals to my camera, I have used all the wiles of a fur trapper. I have paid out as tribute, graham crackers for squirrels, aspen bark for beaver, salt for porcupines, dog food for skunks, cantaloupe rinds for coati, and chicken necks for badgers and foxes. There have been times when the back of my Carryall looked more like a supermarket than a wilderness vehicle.

Enticing birds to the camera is somewhat more difficult, since most exist on specialized diets. I often use feeding tables where an assortment of food may be offered. With convenient perches near the table for the birds to alight upon, it is possible to make photographs with natural backgrounds.

A more rewarding way, however, is locating a nest of young and waiting until the parent birds arrive. Instinct, or mother love, call it what you will, is usually strong enough to make most birds reconsider any hazard. Even with the camera inches away from the nest, few birds will willingly abandon their young.

Photographing subjects at their dens, nests, waterholes, perches, etc., indicates that the camera must be prefocused and set up on the site with a tripod. The approximate size of the subject should be assumed and the camera aimed so this assumed size is centered proportionately in the viewfinder or ground glass. The shutter must be fired by a remote control device. Animals and birds quickly sense that the camera is an inanimate object and soon regard it as part of their environment. In fact, they have often clambered on top of the camera to contentedly eat my offerings. While such incidents seem amusing, they do become frustrating.

The critical problem is setting off the shutter at precisely the right moment. The shutter must be released when the animal assumes the right position. This timing can either make or break the picture. Remember, reaction time displayed by wildlife is considerably faster than with humans. Resetting the shutter after every picture will frighten the creatures for a few minutes, but you will be surprised and pleased to know how quickly they will return.

Implementation of a flash circuit will offer more possibilities. By using a small solenoid which is attached to the shutter the exposure can be made from a distance.

If the circuit is lengthy it is also advisable to use an extra set of batteries in parallel to provide enough electrical surge. In this manner the camera can be operated up to 100 or more feet away without the photographer being at the site, though wires are still needed to carry current.

Exposure is always based on the distance between light and subject matter. For flash work, leaf shutters are a great deal more flexible. They can be synchronized with most shutter speeds. Focal plane shutters are limited concerning synchronization. However, they will stop action with much better facility at slower speeds.

Many 35mm systems can be made more flexible by adding a motorized back, which advances the film and cocks the shutter. By using this procedure, one need never go back to the camera except to change film. There are also 2¼-inch square cameras with the same type of motor control.

A new dimension in remote control photography concerns the use of equipment which includes an electronic eye. When the light beam is broken by the subject, the exposure is made.

I've often been asked how can you be quick enough to photograph animals when you work with long lenses and tripods. The facts are that quality must be maintained, and that speed results from careful camera handling and ''quarterbacking'' the animal's behavior.

Within this framework, first do the technical things automatically. Second, pre-visualize tentative composition.

Prefocusing on a good picture site is an important time-saving approach. When the animals move into your composition, you are then ready to make pictures without wasting time. All these maneuvers give you a tremendous advantage and add to your dexterity of picture composing.

Take advantage of a lull in animal activity to load all film magazines. This is the time to refocus and check the light level. Always watch for clouds. Should they suddenly obscure the sun, you should immediately recalculate the exposure.

All my telephoto lenses use the same camera adapter, which is another time-saving factor. Changing lenses just takes a moment. They all have similar and unique focusing wheels. This makes follow-focusing a simple matter. Such a simple thing as using a tripod with only one control for horizontal and vertical movement increases your maneuverability and speed immensely.

While stalking, I always try to keep the camera on the tripod. I carry the equipment so that the legs are forward. When I stop I am always ready to shoot. If the camera is angled I merely revolve the lens in its mount to level the frame. I rarely ever shoot big wildlife at less than 1/100 sec., usually 1/250 sec. With this median shutter speed I still have leeway to close down the *f*-stop.

The trick is always to be ready, but never to be in a hurry. More pictures are spoiled by needless hurry than can be imagined.

I try to work with the light to the side in order to highlight and create form. The background must in addition present a mood which is so essential to effective portrayals. The *f*-stop should not only control exposure, but should also help to separate the subject from the background.

It is an exciting and exhilarating experience to stalk wild animals. The photographer must constantly study his subjects. He must be able to predict the animal's behavior. Then, at the right moment, when the animal moves into the correct position, the photographer is ready to make pictures.

But there is considerable danger to stalking. I always watch for the animal's reaction to me. To complicate matters more, each individual within the species will have its own peculiar ways. I've discovered that wildlife usually keep a defense perimeter. When the perimenter is penetrated, the animals become nervous. At best they may bolt, and at worst, they will charge. A safe retreat must always be worked out as you shoot and follow the subject.

I'll never forget one of these harrowing escapes, one of the closest we ever had. We were on assignment in Sri Lanka (Ceylon). We had just found a water hole where a mother elephant and her five-year-old were standing. We saw at a glance that there was also a baby elephant lying at the mother's feet. She was gently "rocking" the prostrate form back and forth with her ponderous foot, but the little one didn't stir.

Receding ripples in the water from what appeared to be an old log some distance away indicated what had happened. The little elephant had apparently gone out in the water beyond the safety of its mother. In a flash, his trunk was siezed by a huge crocodile, and the baby had been pulled under. It had drowned before the mother could come to the rescue, and she was still trying to revive her youngster.

Suddenly, she trumpeted and rushed into the bush. Well, I'll tell you, it didn't take long to figure out what was about to happen. We ran headlong for safety just as she came thundering out of the jungle from the opposite direction, trunk up and trumpeting in ear-splitting crescendos. Believe me, there is nothing quite so overwhelming as an elephant bearing down upon you.

As we jumped into the jeep, my light meter thongs became entangled with the legs of the tripod, and the meter spun into the path of the elephant. But we were safe. The driver slammed his foot into the throttle and headed the vehicle for the clearing. Looking back we saw the gray, jungle monolith trampling over the ground where we had just been. Obviously, she was grieving and in a state of shock. Even though our escape had been planned, it was too close to be funny.

I suppose the long and short of all these thoughts on nature photography boil down to how well the image is received by the viewer.

The eye catching, technically well-done photograph always implies that the photographer has never compromised with his medium. In this respect I have always used the camera to portray the highest level of imagery which I have been capable of performing.

I credit this philosophy for my entry into the college teaching field. My credentials led to an appointment as professor of photography at Glendale Community College. My first class consisted of 17 students. Now I have more than one hundred in several classes. Those who have gone into professional photography have done exceedingly well.

Thus, my goal in teaching photography is to make it as meaningful to the student as it has been for me. But better yet, if through the medium of photography, my students see more adroitly, think more clearly, analyze more lucidly, and become more articulate, then it has been a job well done.

Black-chinned hummingbird, Arizona